Innocent Wenceslas TAPSOBA

Artificial insemination of pigs in Burkina Faso

Innocent Wenceslas TAPSOBA

Artificial insemination of pigs in Burkina Faso

ScienciaScripts

Imprint

Cover image: www.ingimage.com

This book is a translation from the original published under ISBN 978-620-6-72395-0.

Publisher:
Sciencia Scripts
is a trademark of
Dodo Books Indian Ocean Ltd. and OmniScriptum S.R.L publishing group

120 High Road, East Finchley, London, N2 9ED, United Kingdom
Str. Armeneasca 28/1, office 1, Chisinau MD-2012, Republic of Moldova, Europe
Printed at: see last page
ISBN: 978-620-8-26095-8

Dedication

TO THE STAKEHOLDERS

OF THE PORK INDUSTRY

OF BURKINA FASO

Acknowledgements

There's an African proverb that says "alone you go faster, together you go further". This memoir has been made possible by a multitude of diverse contributions. We would like to express our gratitude to all those people who, directly or indirectly, gave of their time, advice, energy, knowledge and resources to support us. We would like to say thank you:

- to the chairman of the jury, Mr Albert SOUDRE, Senior Lecturer at the Norbert ZONGO University. It is an honour for us to have you as president of the jury despite your multiple occupations. Your qualities as a man of science leave us in awe. This work gives us the opportunity to benefit once again from your precious and wise advice. You have our deepest respect;

- to Mr. Boureima TRAORE, Assistant Professor, Centre Universitaire de Dori / Université Thomas SANKARA, member of the jury, please find herewith the expression of our respect and deep gratitude;

- Mr Moussa ZONGO, Full Professor at Joseph KI-ZERBO University, our dissertation supervisor and member of the jury. Your scientific and pedagogical qualities have always greatly impressed us. Please find here the expression of our deep respect and admiration for your scientific rigour;

- to Mr Dominique OUEDRAOGO, Assistant Professor at the Centre Universitaire de Ziniaré / Université Joseph KI-ZERBO, our dissertation co-director and member of the jury, it is a great honour for us to have you as a judge of this modest work. Please find here the expression of our deepest gratitude.

We would also like to thank the hard-working lecturers at UFR/SVT and particularly those in the Animal Biotechnology Masters;

We would like to thank the Projet Régional d'Appui au Pastoralisme au Sahel, phase 2 (PRAPS-2) for its financial support;

We would like to thank the Projet d'Appui au Développement du secteur de l'Elevage au Burkina (PADEL-B) which, through its structuring investments, has enabled the construction of a glasshouse at the Loumbila breeding station, the construction and equipping of the pig laboratory and the acquisition of purebred exotic boars;

We would like to thank the technical and support staff of the national centre for the multiplication of efficient animals, particularly Mr Mamadou DIALLO and Mr Noufou COULIBALY, for their assistance in collecting the data;

We would like to thank the pig farmers of Burkina Faso for their support for this study, and our fellow students in Animal Biotechnology from the class of 2021-2022 for their cooperation.

Finally, we would like to thank our entire family for all the support they have given us.

Table of contents

Dedication ... 1
Acknowledgements ... 2
Summary ... 5
INTRODUCTION ... 6
PART I: LITERATURE REVIEW ... 8
Chapter I: General information on pig farming ... 9
Chapter II: Boar seed technology ... 21
Chapter III: Artificial insemination (AI) in pigs ... 26
PART II: EXPERIMENTAL STUDY ... 29
Chapter I: Materials and methods ... 30
Chapter II: Results and discussion ... 38
CONCLUSION AND OUTLOOK ... 45
Bibliography ... 46

Summary

Demand for pigmeat in Burkina Faso is growing all the time, against a backdrop of low productivity among existing breeds. The development of pig production requires the use of animal biotechnologies. The aim of this study is to help improve the productivity of the pig species in Burkina Faso. Pig semen technology was developed at the Loumbila breeding station. The methodology consisted in training and collecting semen from 08 pure-bred boars introduced to the station. The semen was subjected to macroscopic and microscopic analysis in order to assess its quality and guide decision-making on its conservation and use in artificial insemination. The number of ejaculates collected was 33, with a total volume of 4296.5 ml from 5 boars aged [15 - 16] months that had repressed favourably. The reaction time to jumping was 3.13 minutes for the large white breed, 3.08 minutes for the landrace and 4.09 minutes for the duroc. The average duration of ejaculation was 5.59 minutes, 5.75 minutes and 5.00 minutes respectively for the large white, landrace and duroc. The average volumes of ejaculate obtained were 141.16 ml, 136.67 ml and 191.08 ml for the large white, duroc and landrace breeds respectively. The colour was creamy white. Sperm motility averaged 2.99 on a scale of 0 to 5. Mean sperm concentrations were 195.45 x 10^6 , 198.59 x 10^6 , 383.01 x 10^6 for landrace, large white, and duroc respectively. The average diluent volume was 1.03 ml, 1.22 ml and 1.78 ml for landrace, large white and duroc respectively. Individual motility ranged from 20 to 60%. Average dose production was 12, 11 and 20 for landrace, large white and duroc respectively. Doses were stored at 15-18°C. Cervical inseminations were performed on 145 sows on 25 farms in the Ouagadougou and Bobo-Dioulasso areas. Pregnancy was diagnosed by ultrasound 35 days later on all the inseminated sows. The AI success rate was 35.42% (n =144). The average gestation period was 113.77±2.66 days. Duroc boars performed better than Large White and Landrace boars. The Piétrain breed seems to have had difficulty adapting to the negative results of the two boars.

Key words: artificial insemination, pigs, Burkina Faso, seeds, production, packaging.

INTRODUCTION

Burkina Faso is an agropastoral country, and livestock farming plays an important role in terms of the number and diversity of its livestock. With more than 9,939,786 head of cattle, 22,084,362 head of small ruminants, 1,454,087 head of pigs, 36,275,347 head of poultry and many other species such as donkeys, horses and camels (MARAH, 2021), livestock farming employs more than 86% of the working population and contributes 18% of the gross domestic product (GDP) (UNDP, 2011).

Pig production, once practised by a minority of the population, has been growing rapidly for nearly three decades (Kiendrebeogo, 2006). The country's pig herd represents extraordinary meat production potential and a source of income for producers and intermediaries working in the pork sector (Djiguimde, 2019). It is increasingly becoming a favourite national occupation and a real tool for creating jobs, empowering vulnerable people, fighting poverty and helping to achieve food and nutritional security.

Eight (08) pig breeds are found in Burkina Faso. These are the local breed, the Korhogo, the Large White, the Landrace, the Pedestrian, the Duroc, the Hampshire and the Berkshire (MRA, 2003). Characterised by its short reproduction and production cycle, high feed efficiency and good adaptation to different ecosystems, the pig is firmly established as the animal of choice in livestock development policy in our countries (Muys et *al,* 2003). These reasons explain why various African countries have started importing pigs of improved breeds in order to improve the production of local breeds. Artificial insemination (AI) is the preferred breeding method in the majority of intensive pig production systems worldwide (Rodriguez et *al.,* 2016). Porcine artificial insemination is not highly developed in West Africa (Dotché *et al.,* 2019). However, this technology is an alternative solution to the sexual dimorphism observed in certain animal species, in this case boars, which are more often than not faced with natural mating difficulties (Knox, 2016). It also helps to reduce labour time and protect animal health by avoiding venereal infections (Knox, 2016). To make better use of AI, efforts must be made to produce quality doses in other animal species, especially pigs (Dotché et *al.,* 2019). Reproduction by natural mating is almost exclusively practised on small farms where the sow population does not exceed 20 units. Compared with AI, this involves more handling of the animals, and therefore more time and greater risk of accidents (Feller *al.,* 2004). Artificial insemination remains the biotechnological tool that ensures the rapid and safe dissemination of high-performance genetic material, which facilitates the genetic improvement of farm animals, including pigs (Bitty ,2014). The use of semen technology would reduce the costs and risks associated with importing live boars. The use of locally produced semen would reduce the costs of the technology (the costs of importing and preserving imported semen) (Moussa ,2018). Work on these exotic breeds has only been done in Europe (Bitty ,2014). But what about the characteristics of the seed of these animals in our climates and rearing conditions?

The overall aim of this study is to help improve pig production by developing low-cost semen and artificial insemination technology.

Specifically, this involved :

(i) evaluate the characteristics of exotic boar semen introduced into Burkina Faso,

(ii) determine the fertility of conditioned semen when artificially inseminated.

Our paper is divided into two main sections: the literature review, which summarises the general background to pig breeding, discusses boar semen collection and evaluation, and porcine artificial insemination; and the experimental study, which deals with the study equipment and methods, and presents the results and discussion.

PART I: LITERATURE REVIEW

Chapter I: General information on pig farming

I. General

Pigs are reared all over the world for their meat. With global production estimated at 110.5 million tonnes in 2022 (OECD/FAO, 2022), pork ranks second in the world after poultry, followed by beef.

Pig farming is growing rapidly in the various regions of Africa. The most popular pig breed is the local pig, which requires few inputs and is highly adaptable to difficult farming conditions. Pig farming is therefore compatible with the socio-economic conditions of poor populations (Agbokounou, 2016).

The characteristics of the traditional livestock farming system are sometimes found in tropical Asia and America (Nonfon, 2005). It is often village-based and extensive. However, this system can be found in urban and peri-urban areas where the local breed, improved European and North American breeds (Large White, Landrace, Duroc, Piétrain, Hampshire) or hybrids of these breeds crossed with the local breed or products of several uncontrolled crosses are bred (Lukusa, 2005). In the case of breeding improved breeds and their crosses, breeders are improving the extensive form, which sometimes tends towards a semi-intensive and sometimes intensive form.

II. Taxonomic description

The pig is an omnivorous domestic mammal. According to Linnaeus (1758), the pig belongs to the animal kingdom, the chordate phylum, the vertebrate subphylum, the mammalian class, the therian subclass, the eutherian subclass, the ungulate order, the suidae family, the suidae subfamily, the genus Suis, the species sus scrofa and the subspecies domesticus.

It is closely related to the wild boar, with which it may interbreed. The adult female is the sow, the young female bred for reproduction is a gilt, the male is the boar, the young male is the verrason and the young pig (before weaning) is called piglet, piglet, goret (or suckling pig on the plate), the young weaned pig is called nourrain (or nourrin). The term also refers to the meat provided by this animal.

III. History of pig farming in Burkina Faso

Pig farming has a religious connotation in Burkina Faso. The majority of pig farmers are Christians and animists. Several accounts place the introduction of pork around the 1920s, thanks to missionaries and colonial administrators (MRA, 2003).

Initially, the aim was to satisfy their own eating habits. Later, in order to win the esteem of the people, it was popularised among the population, through catechists who were the relays between the priests and the people as part of a process of raising awareness and developing the Christian faith in rural areas (FAO, 2012). Some traditional chiefs, who quickly converted to Catholicism, played a major role in the development of pork, particularly around Réo and Tenkodogo (MRA, 2003).

The pig does not seem to be an animal known "to the ancestors" in several Burkinabe social groups. Nowhere has it been reported that the pig, like other species such as poultry (chickens, guinea fowl), small ruminants (sheep and goats) or even dogs, has been the subject of specific use in customary and religious rites (FAO, 2012).

With anarchic cross-breeding methods and the absence of a selection and genetic improvement scheme, the animals have rapidly lost certain production performances over the years, mainly through successive inbreeding (MRA, 2003).

IV. Pig herd in Burkina Faso

Pigs are one of the five priority species reared in Burkina Faso, but according to the country's statistics, the industry is still evolving. The pig population was estimated at 660,582 in 1989, 1,889,234 in 2003 and 1,321,464 in 2018, according to the first national livestock survey (MARAH, 2021).

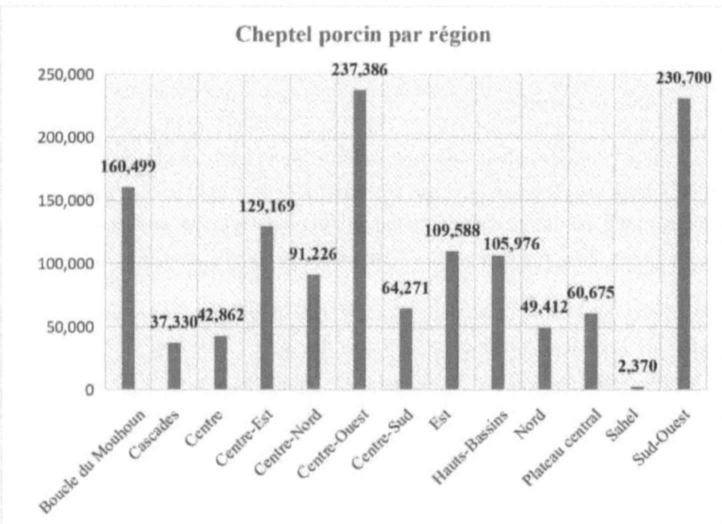

Figure 1: Distribution of pigs in Burkina Faso (MARAH, 2021)

Three regions alone account for 47.57% of the national workforce. These are the Boucle du Mouhoun region with 12.14%, the South-West with 17.45% and the Centre-West with 17.96%.

V. Livestock systems and pig production

Two main systems coexist, as do the breeding of other species in the country. These are the traditional system and the "improved" system (CIRAD, 2006).

At national level, the traditional system dominates in 80% of cases and is characterised by pigs roaming in search of food, the absence of care, the occasional distribution of dolo and drinking water, which enables the owner to control the number of head (births, losses, remaining numbers) that he owns. The animals are kept in seasonal confinement (tethered or in traditional housing) during the wintering period (FAO, 2012).

The confinement farming system, considered to be improved, is divided into two sub-systems: the improved system and the modern system, which uses exotic animals, often pure breeds. These farms benefit from greater investment and monitoring.

According to the FAO (2012), pork production in Burkina is growing rapidly. Its production is estimated at 39,800 tonnes of meat in 2022 (Porphyre, et *al.*, 2022).

In addition to production, which increases every year, consumption follows the same trend. In Burkina Faso, pork is eaten in a variety of ways. The best-known preparation is "oven-baked pork". Other forms of cooking, such as "braised pork", "pork soup" and sometimes on skewers, are also on the rise.

It is marketed either as live animals, carcass meat or in processed forms.

VI. Genetic improvement in pig farming in Burkina Faso

Genetic improvement for a domestic species is the set of methods and techniques (cross-breeding, selection, genetic engineering, etc.) aimed at replacing one population with another which may or may not be derived, in whole or in part, from the previous one and which is better adapted to the economic objectives being pursued (Annick, 1995).

The development of improved pig farms based on a standard diet has led to the expansion of exotic breeds selected for a better feed conversion ratio (FCR), higher muscle yields, less fat cover and larger litter sizes (IFIP, 2013). Uncontrolled breeding and cross-breeding are a significant threat of genetic erosion for the pig species in Burkina Faso.

VII. Constraints on pig production

The constraints of pig production are universal because they always revolve around the same issues, namely how to improve :

- growth rate, which is assessed on the basis of the GMQ for fattening;

- feed efficiency, which is assessed on the basis of fattening CI;
- body composition, which can be seen in carcass yield, meat content, etc.

and muscle content.

In addition to these challenges of improving production traits, there are also selection challenges such as :

- age ;
- fertility rate ;
- the number of piglets weaned per sow per litter, which reflects the sow's maternal quality;
- the number of weaned piglets born to the sow, which reflects the vitality of the piglets;
- the number of functional teats the sow has.

In Burkina Faso, access to pigs that perform well in both reproduction and production, in terms of availability and cost, remains a major constraint, in addition to food and health issues.

VIII. Anatomy and physiology of boar reproduction

The boar's reproductive system, shown schematically in Figure 2, includes the testes, epididymides and adnexal glands. The boar is characterised by the large size of its reproductive organs (testes, seminal vesicles and bulbourethral glands) compared with other domestic mammals.

The production of sperm and sex hormones takes place in the testicles. They are protected from temperature fluctuations by the scrotum, a membrane that covers the testicles and regulates temperature to ensure that the testicles remain at a more or less constant temperature. Extreme temperatures such as heat or fever have a negative impact on the quantity and quality of sperm produced (IFIP, 2013).

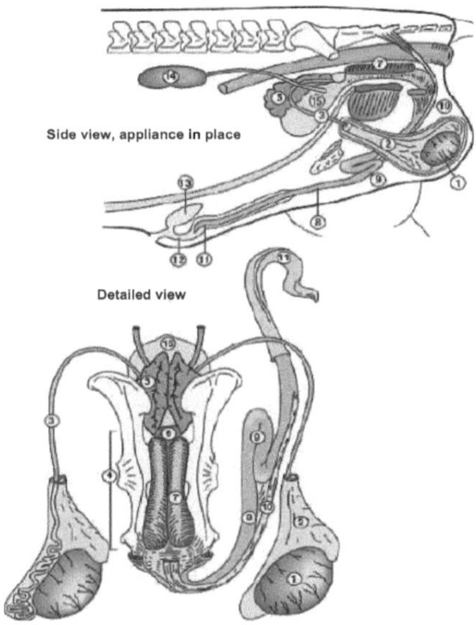

1. **Testis and its envelopes**
2. **Epididymis**
3. **Deferential canal**
4. **Pelvic bone**
5. **Seminal vesicle**
6. **Prostate**
7. **Cowper's glands**
8. **Penis**
9. **"penile "s**
10. **Penis retractor muscle**
11. **Spiral end**
12. **Sleeve**
13. **Preputial glands**
14. **Kidney**
15. **Bladder**

Figure 2: Reproductive organs of the boar (IFIP, 2013)

VIII. 1. Anatomy of the testicle

The testicles are paired organs with a dual function:

> the gametogenic or spermatogenic function, which can be likened to a genuine exocrine function and is carried out by the seminiferous tubules;

> the endocrine function of secreting male hormones, androgens, provided by the Leydig cells located between the seminiferous tubules. The size of the testicle varies according to the animal species.

The ratio of testicle weight to body weight is very high in boars, at 4.2.^{103} . The testicles are relatively large in relation to the size of the animal, with the left testicle generally being larger than the right. The average weight of the two testicles is between 300 and 800 grams. They are elliptical in shape in boars and are located high under the anus, in the perineum; they are generally not very detached from the body but may become a little more so with age (Martin, 1993). The testicular parenchyma is grey to burgundy in colour and darker with age (Ladds, 1996).

VIII. 1.1. Intra-testicular structures: seminiferous tubules and interstitial spaces

The testicular parenchyma is grey to dark burgundy in the boar, with a prominent mediastinum

(Ladds, 1996). The parenchyma is composed of seminiferous tubules and interstitial spaces and divided into lobes and lobules by connective trabeculae derived from the albuginea and rete testis (Figure 3).

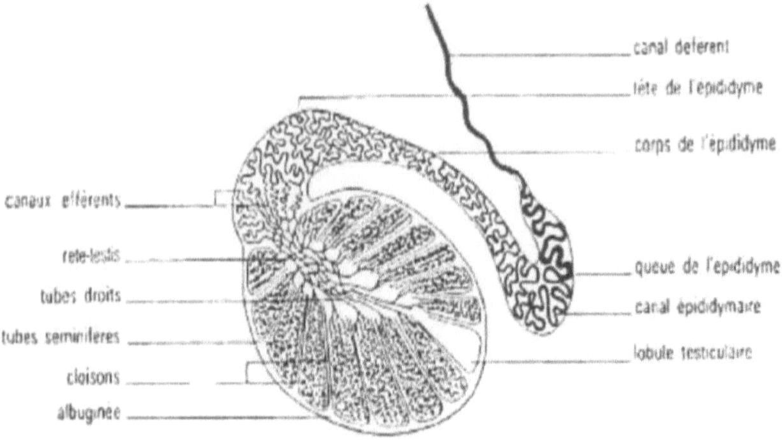

Figure 3: Internal structure of the testis (INRAP)

VIII. 1.2. Testicular function: spermatogenesis

Spermatogenesis is the entire process of multiplication and cellular differentiation of the cells of the male germ line which, starting from the stem cells or spermatogonia, leads to the production of spermatozoa. This is a highly productive process that takes place continuously in the seminiferous tubules from puberty onwards. Spermatogenesis is maintained by hormonal regulation. GnRH stimulates pituitary production of LH and FSH. Spermatogenesis is stimulated directly by the action of FSH on Sertoli cells and indirectly by the production of testosterone resulting from the action of LH on Leydig cells. Negative feedback is exerted by inhibin and testosterone produced by the testis. FSH and testosterone are essential for maintaining spermatogenesis and libido in adult boars.

VIII.2. spermatogenesis

VIII.2.1. The epididymis: maturation and storage of spermatozoa

The epididymis is an organ adjacent to the testis that measures 15 to 20 cm in length in boars and contains an elongated, highly contoured tube that connects the efferent ducts from the testis to the vas deferens, measuring 70 cm in boars (Dacheux and Paquignon, 1987). It is bulky, weighing an average of 80 grams, with extremes ranging from 46g to 110g (Barone, 2000). From an anatomical point of view, it is usually separated into three distinct anatomical parts: the head, which is ventral and folded over the capitate end of the testicle, the body and the tail, which is larger than the head and well detached from the testicle. The physiology of the epididymis is more complex, as there are no anatomical landmarks to distinguish the different epididymal regions, which are specialised in specific activities.

VIII.2.2. Epididymal functions

The epididymal epithelium reabsorbs, secretes and concentrates substances that help modify the composition of the epididymal fluid to promote the survival of mature spermatozoa before they are released into the tail of the epididymis. This epithelium also acts as a barrier between the blood and the lumen of the tube, providing a favourable environment for the survival and maturation of male gametes (acid pH, low electrolyte levels, high K+ levels, absence of fructose) (Dacheux and Paquignon, 1987).

VIII.2.3. The adnexal glands and the production of seminal plasma and "tapioca"

VIII.2.3.1. The seminal vesicles and the prostate: production of seminal plasma

The seminal vesicles are reduced until puberty, but become voluminous in adult boars. They develop under the effect of testosterone, more or less early depending on the breed, and increase from a few grams at birth to an adult weight of 500 to 800 g. Positioned dorsally to the bladder, they lean against each other in the median plane. In boars, the two glands have a common connective capsule and a fine muscular tunic. The interlobular septa consist mainly of connective tissue and a few small muscle cells. The tubular lumens are wide and the secretory epithelium is wrinkled. The main cells secrete an abundant viscous fluid which makes up 10 to 30% of the ejaculate (Wrobel, 1990). This fluid is very rich in calcium and magnesium and contains energy substrates for the spermatozoa, such as glyceryl phosphoryl choline and ergothioneine in boars (it is known that inositol cannot be used by the spermatozoa as a metabolic substrate, even though it is present in very large quantities), as well as proteins in very high concentrations (100 mg/ml), structural proteins as well as enzymes and enzyme inhibitors... The role of secretions from the seminal vesicles is to activate the sperm's ability to move and to provide them with the nutrients they need to survive (Flowers, 1996). The prostate is relatively underdeveloped in boars and is located cranial to the seminal vesicles (Flowers, 1996). It is a well-irrigated, compact gland whose secretions, which make up 30 to 60% of the total volume of the ejaculate and are rich in calcium, zinc, buffer substances, bacteriostatic amines, as well as steroids and oestrogens in boars, are difficult to collect. The role of prostatic secretion is to neutralise seminal plasma, which has been rendered acidic by an accumulation of bicarbonates and lactates, and to initiate the active movement of ejaculated spermatozoa (Wrobel, 1990).

VIII.2.3.2. Other adnexal glands and their secretions

Other adnexal glands include the urethral glands and preputial glands: The urethral glands or Littré's glands are small mucous glands which open into the penile urethra via a canaliculus. They secrete during coitus and their secretion is the first to come out during ejaculation; it is small in volume and clear serous in appearance. They are also thought to have the function of cleaning the duct of urine remains before the ejaculate passes through. The preputial glands or

Tyson's glands are located near the preputial orifice, where the remains of urine accumulate and decompose, giving the unpleasant odour characteristic of boars. Their secretion is a source of pheromones, vehicles of the male effect in sexual interactions, but it is not part of the ejaculate.

The boar's penis is long, slender and fibro-elastic, measuring around 60 cm long and 2 cm thick. The tip of the penis has only a rudimentary glans penis and is twisted around itself. This torsion is accentuated during erection (Bitty, 2014).

IX. Anatomy and physiology of sow reproduction

The sow's reproductive system comprises the gonads, genital tract and accessory glands.

IX.1. The gonads

There are two small ovaries (Figure 4). Their appearance changes according to the physiological state of the sow. Each ovary is made up of 3 types of tissue: the albuginea (the fibrous membrane covering the ovary), the medullary zone (the central part of the ovary made up of blood vessels and nerves) and the cortical zone.

The ovary is the site of cyclical and periodic follicle budding. The follicles have different names depending on their stage of development: the primary follicle (in definite number and in stock from the birth of the female), the secondary follicle, the cavitary follicle (formation of the antrum filled with follicular fluid) and the mature follicle or De Graaf follicle (release of an oocyte by bursting of the albuginea). The follicle transforms into a corpus luteum from the lutein cells, which produces progesterone and will persist if fertilisation occurs (Feller et *al.*, 2004).

IX.2. The genital tract

- **The oviducts** in sows measure 20-23 cm and are very flexuous (Figure 4). They capture oocytes II and are the site of fertilisation at their upper third (%) (Bitty, 2014). The oviducts are also known as uterine tubes or Fallopian tubes (Feller et *al.*, 2004).

- **The uterine horns**, also known as the uterus or matrix, come together to form the body of the uterus (Figure 4). The wall of the uterus is made up of 2 tissues (the endometrium and the myometrium) and its development depends on the physiological state of the sow. In pigs, sperm deposition is therefore intrauterine. (Feller et *al.*, 2004).

- **The cervix**: this is a muscle measuring 10-23 cm long and 2-3 cm wide which separates the uterus from the vagina (Bitty, 2014) (Figure 4). It opens only during heat and parturition.

- **The vagina**: this is a muscular-membranous canal located between the vulva and the cervix and has a smooth, muscle-free structure. It is the female's mating organ and is around 20 cm long (Figure 4). The vagina has thick walls that taper in the caudal part near the uterus. The vestibule of the vagina includes the urinary meatus, which forms the exit of the urethra. The hymen is a thin membrane that separates the vagina from the vulva. It is torn at first mating. (Feller et *al.,* 2004).

> **The vulva**: this is a urogenital duct made up of vulval lips, an appendix and the clitoris. It is triangular in shape (Feller et *al*., 2004).

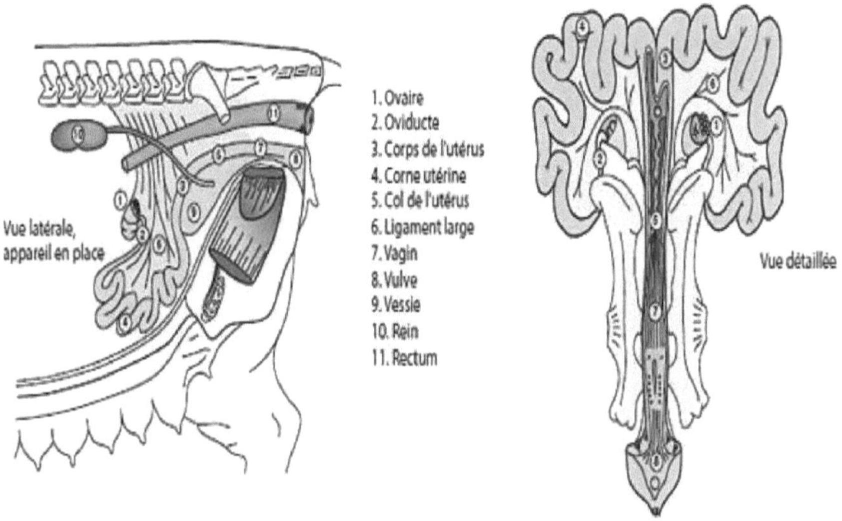

Figure 4: Genital tract of the sow (IFIP, 2013)

IX.3. The mammary gland

It is made up of 2 or 4 or more independent 10-12 quarters supported by a transverse suspensory ligament which separates the front from the back and a longitudinal suspensory ligament which separates the right from the left and which ascends towards the anus. It is made up of mammary tissues, including connective tissue, which is the fleshy part that supports and fills the udder, glandular tissue, which secretes milk, and blood vessels, nerves and muscles (sphincter). Complete growth of the udder only occurs at the end of gestation and during lactation (Vincent, 2014).

X. Estrous cycle

The sow is a continuous poly-oestrous animal with spontaneous ovulation. The age at puberty varies between 6 and 8 months and the first farrowing can occur at 12 months of age, sometimes earlier (Da, 2004). The weight of the animal at puberty is between 90 and 120 kg (ITP, 1991).

The oestrous cycle lasts an average of 21 days (Vaissaire, 1977; ITP, 1986). It is represented by two phases: a follicular phase comprising pro-estrus and estrus, and a luteal phase comprising metestrus and diestrus (Figure 4).

X.1. Follicular phase

> **Pro-estrus** :

It is characterised by a decrease in the corpus luteum from the previous cycle. This is the period of follicular maturation or preparation for the rut. There is a considerable increase in oestrogen secretions, which peak at around 85 pg/mL of blood (Da, 2004).

> **Oestrus :**

Oestrus is the Greek word for heat, due to the particular behaviour of the female during this period. It is characterised by the presence of mature follicles in the ovary, endometrial hyperaemia and ovulation, which corresponds to the end of follicular maturation and its rupture to release the ovum. Oestrus is therefore the period of ovulation, but also the start of corpus luteum formation (Figure 4), which is manifested by a change in the female's behaviour (restlessness, reduced or loss of appetite, pointed ears, tail wagging, swollen and reddened vulva (proestrus), vocalisation/growling). Heat corresponds to the male's acceptance period. These signs can be used to identify the ovulation period for a controlled natural mating. All these changes are due to oestrogen (Toussaint, 2000).

X.2. Luteal phase

This is the phase of formation of the corpus luteum and all the organic changes are linked to progesterone. Behaviourally, during this phase, the sow no longer accepts the boar. There are two stages to this phase:

> **Metestrus** :

It begins after ovulation and lasts 13-14 days in cycled females. This is the period of activity of the corpus luteum, the ova are 4-5 mm in size and the ovulation peak is visible with little luteal tissue (Figure 5). It is linked to the anabolic phase of the corpus luteum, which is characterised in the uterus by changes in preparation for embryo implantation.

> **Diestrus** :

This is the sow's sexual resting phase, followed by a considerable drop in progesterone levels around day 15ème or day 16ème . This drop corresponds to a rise in plasma levels of prostaglandin F2a (PGF2a), which leads to a regression of corpora lutea. This is followed by an increase in oestrogen production, which reaches its peak, and another cycle resumes with pro-oestrus (ITP, 1991). High levels of oestrogen in plasma induce the oestrus phenomenon (Bories, 2009).

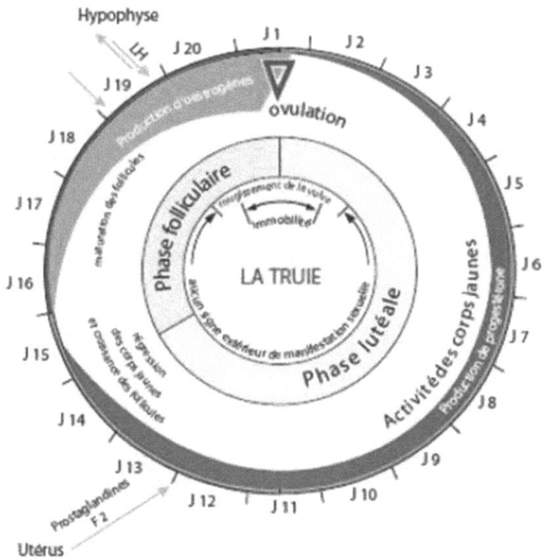

Figure 5: Physiological changes associated with sexual cycles in sows (IFIP, 2013)

X.3. Hormonal regulation of the sexual cycle

The sow's oestrous cycle is controlled by the hypothalamus-pituitary complex (Figure 6). GnRH, secreted pulsatile by the hypothalamus, induces FSH and LH secretion by the pituitary gland (Da, 2004). FSH promotes follicular growth and enables oestrogen to be conserved. LH completes follicular maturation, triggers ovulation and secretes oestrogen (Brassard et *al.,* 1997). This is regulated by several events grouped into two phases surrounding ovulation. During the pre-ovulatory phase, oestrogen produced by the follicles has a positive feedback effect on FSH and GnRH.

(Figure 6). Oestrogen stimulates the activity of the hypothalamus by positive feedback, which secretes pulsatile GnRH that acts on the gonadotropic cells of the pituitary, triggering massive production of FSH and LH. The oestradiol produced by the follicles can directly stimulate the gonadotropic cells of the pituitary gland by means of a short positive feedback loop to trigger the production of FSH and LH. The very high level of LH produced triggers ovulation and then causes the mature follicle to transform into a corpus luteum. During the post-ovulatory phase, the corpus luteum transforms, grows and secretes progesterone, which acts on the peripheral cells (endometrium, myometrium, cervix) and inhibits the release of GnRH by negative feedback to the hypothalamus (Figure 6). Progesterone can also act directly on the pituitary gland via negative feedback to inhibit any massive release of gonadotropins from the pituitary gland (Figure 6). At the end of the luteal phase, in the absence of an embryo, the uterus secretes prostaglandin, which induces luteolysis of the corpus luteum (Da, 2004).

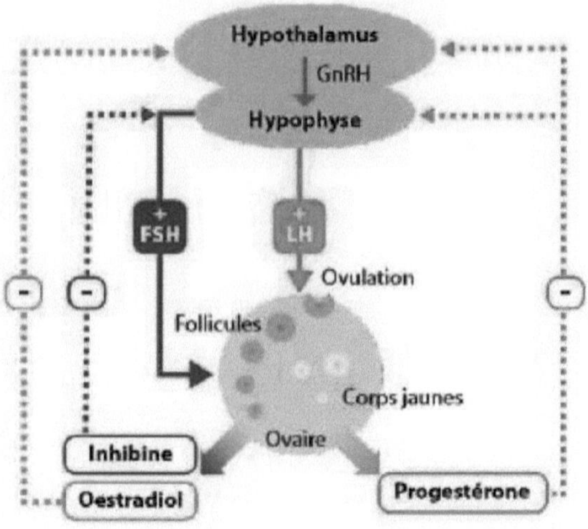

Figure 6: Hormonal function in sows (IFIP, 2013)

Chapter II: Boar seed technology

I. General information on animal semen

In boars, semen is a clear, viscous liquid produced during coitus by the genitals and adnexal glands. It contains both spermatozoa and secretions from the adnexal glands known as seminal fluid. Spermatozoa are haploid cells that carry the male genetic heritage. They are elongated and mobile, and must be able to reach and fertilise the oocyte in the female genital tract. The seminal fluid contains the nutrients necessary for the survival and movement of the spermatozoa once released during ejaculation. The semen can be collected and its quality assessed.

II. Harvest

Puberty in the young boar is reached between 4 and 5 months of age, but it is around 10 to 12 months that semen production reaches its optimum. In the last week of foetal life, the testicles descend into the scrotum; at 10 weeks of age, spermatozoa begin to form in the seminiferous tubules; at 16 weeks of age, the young boar acquires the ability to erect; At 20 weeks of age, spermatozoa are present in the epididymis, representing sexual maturity; at 22 weeks of age, spermatozoa are present in the semen and at 12 months of age, maximum fertility is reached (Memento, 2013).

The boar is one of the animals that ejaculate abundantly. In a single ejaculation, an adult boar can produce between 100 and 600 ml of semen (Rillo, 1996) with a hydrogen potential (pH) of between 7.3 and 7.9. There are several harvesting techniques:

II.1. Harvesting by hand

Boar semen can be harvested manually, without the need for an artificial vagina as in the bovine species. The process takes an average of 4 to 10 minutes, depending on the age of the boar and the frequency of harvesting (Feller, 2005). Both the boar and the operator need to be comfortable and secure during this period. Young males are trained to jump on a dummy. In addition, the boar can be stimulated using urine from sows in heat, semen from another boar or an injection of commercial pheromone. Once the boar has made its first good jump, training is repeated every 3 to 4 days to familiarise it with the collection method after 3 weeks. Boar semen is collected at intervals of 5 to 8 days. A very short interval reduces the sperm concentration in the ejaculate, whereas a very long interval between consecutive collections increases the percentage of agglutination of the ejaculate (Le Roux, 2002).

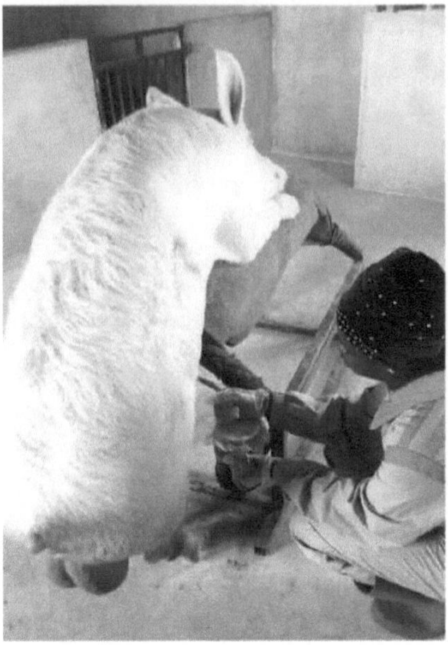

Figure 7: Manual collection of boar semen (Gakou, 2018)

II.2. Semi-automated harvesting

This approach has been developed to improve the productivity and quality of the boar semen harvested. The technique makes it possible to harmonise the harvest regardless of the operator and the number of boars harvested per day and per operator, according to a guide document from IMV Technology.

The penis is held in place by an artificial vagina containing a cone and an ejaculate collection filter. The pre-putial fluid is manually discharged by the technician using two gloves. The boar's penis is connected to the artificial vagina linked to the dummy. When ejaculation is complete, the boar detaches itself from the artificial vagina itself and collection is complete (IMV Technology guide).

II.3. Automated harvesting

This is a more advanced version of the semi-automatic collection system. This technique produces air pressure and stimulates the boar even more, giving the same comfort as natural mating.

This technique is suitable for large-scale seed collection centres.

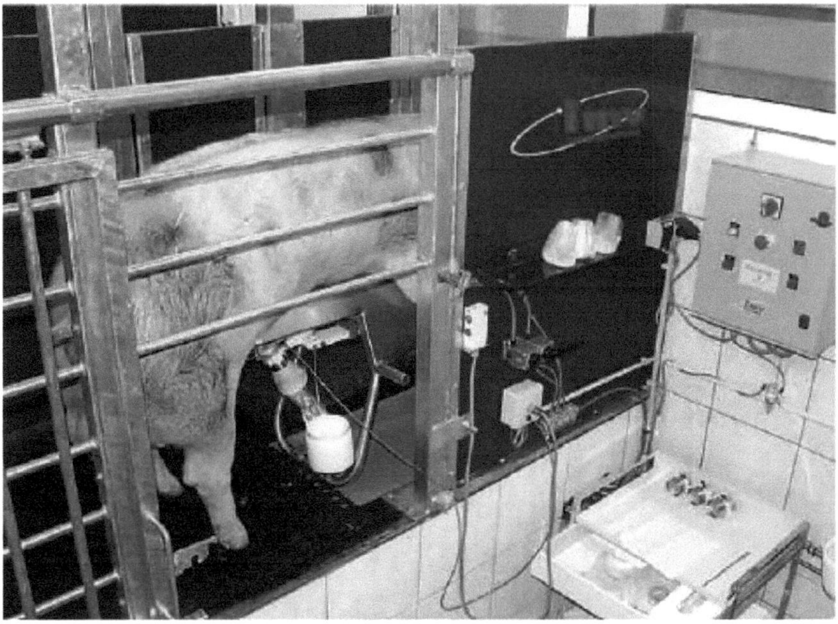

Figure 8: Automated semen collection (Guide, IMV Technologie)

III. Seed quality assessment

III.1. Visual and olfactory assessment and quantification of semen

Once the harvest is complete, the filter is removed with the tapioca and the seed can be enjoyed:

- its colour: creamy white, traces of blood, presence of urine;
- its smell: odour of urine, infection ;
- its volume: this requires a high-precision balance to give the volume (in ml) of the ejaculate collected.

III.2. Microscopic evaluation

This assessment consists of :

- observe the motility (liveliness) of the spermatozoa under the microscope, check for agglutination and assess the percentage of dead spermatozoa (or % alive);
- use a spectrophotometer to determine the sperm concentration of

the collected ejaculate (figure 9).

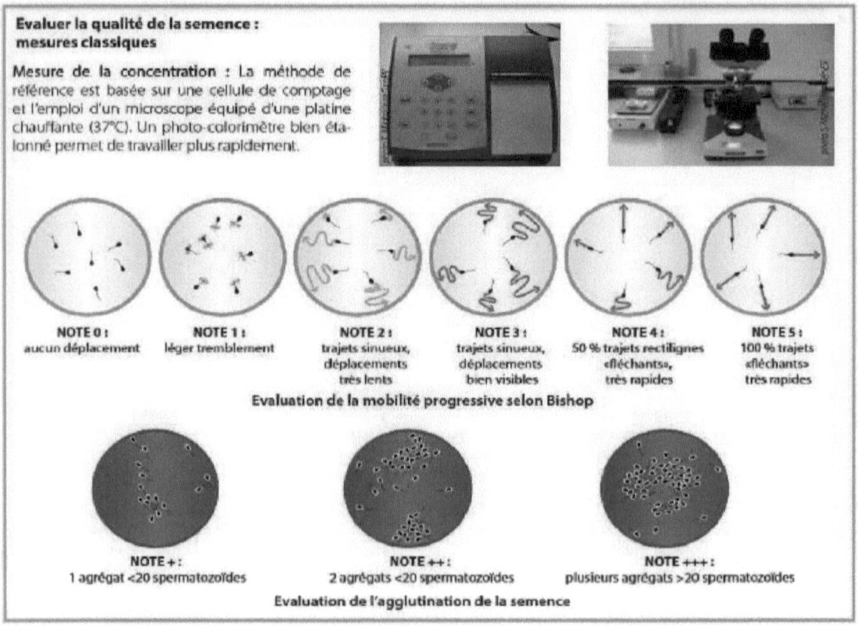

Figure 9: Seed quality assessment (IFIP, 2013)

IV. Seed dilution and conservation

VI.1 Dilution

The major constraints of artificial insemination techniques are twofold: the sexual physiology of the sow on the one hand, and the 'in vitro' conservation of boar semen after dilution on the other (Mesnil, 1956).

Dilutors must take into account a certain number of physico-chemical imperatives such as (i) having an isotonic osmotic pressure; (ii) containing buffer substances capable of maintaining the pH close to neutral, containing substances necessary for the metabolism of the spermatozoon; (iii) being free of bacterial agents.

Thinners are generally made up of :

- Glucose, which is better assimilated by spermatozoa than fructose, is the best source of energy. It also helps maintain the osmotic balance of the environment.
- Sodium citrate, sodium bicarbonate and Tris (trihydroxymethylaminomethane) are the

buffers most commonly used for boar semen.

- EDTA, contained in most diluents, is thought to stabilise cell volume (Bredderman, 1971). In addition, its chelating properties prevent the peroxidation of membrane phospholipids, which is detrimental to the maintenance of fertility (Shanon et al, 1992).

All these elements are dissolved in distilled water. To assess the effectiveness of the different diluents, the percentage of motile spermatozoa and their fertilising power are traditionally determined after storage for a given period in the diluent in question.

VI.2 Conservation

The insemination trials attempted by ITO, NIWA and KUDO (1949) achieved 34.7% fertilisation with sperm kept pure for 3 days. After collection, spermatozoa mixed with seminal plasma alone have a short lifespan. The incorporation of a medium likely to prolong their survival and maintain their fertility is therefore essential (Paquignon, 1988). A storage temperature of between +15°C and +18°C has generally been suggested as the best way of preserving semen. However, it is difficult to maintain this temperature when the semen is sent to the breeder. The doses are then subject to external temperature variations and the quality of the semen is affected.

Chapter III: Artificial insemination (AI) in pigs

I. General information on swine AI

Artificial insemination is used to create and disseminate genetic progress, but also as a tool for controlling the reproduction process. It is now widely used by breeders in North America, Europe and Asia. In France, it is mainly carried out using semen produced in a collection centre (79%) combined with semen collected on the farm (4%) and, more rarely, natural semen (4%) (Boulot, 2008).

II. Heat detection

Oestrus can be detected through changes in behaviour and the morphology of certain organs due to hormonal secretions (Hanzen, 2009). Among the main signs of oestrus, the immobility reflex is "the sign" (figure 10). But there are other signs which, to a greater or lesser extent, are also signs that the sow is in heat:

- Enlargement of the vulva during all periods of heat;
- the presence of sticky mucus on the vulva, which is more prevalent at the end of heat;
- attempts to climb on other sows, especially when the sows are in a group;
- inappetence due to the increase in oestrogen hormones, which have a hunger-suppressing effect on the body;
- long-lasting grunting;
- ears pricked upwards ;
- tail up.

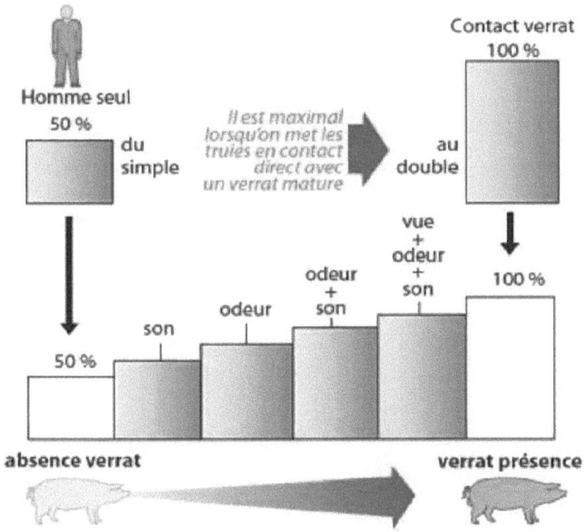

Figure 10: Sow immobilisation rate (IFIP, 2013)

III. insemination

Today, the relevance and use of AI as a reproductive tool is well established in most Western countries. Indeed, the first scientific article on artificial insemination was published in 1897 in Cambridge (Heape, 1900), but it was established as a reproductive biotechnology tool following the high demand for new horses by the Russian army in the 1920s. However, in-depth research began in the 1930s with the work of McKenzie in Missouri (USA). He was the first to manufacture an artificial vagina for collecting boar semen. In Europe, it was Milovanov in 1938 and Polge in 1956 who trained boars to jump on a dummy for semen collection.

The first doses of boar semen were developed in Russia and were based on a glucose solution with sodium and potassium tartrate or sodium sulphate and peptone. The first storage temperatures were between 7 and 12°C, but Ito et *al* (1948) quoted by Du Mesnil, (1956) recommend a storage temperature between 15° and 20°C to ensure good fertility.

Following the success of freezing bull semen, several trials were carried out to freeze boar semen, but fertility after insemination remains low compared with the use of fresh semen.

Artificial insemination has a number of advantages:

- several sows can be inseminated with a single ejaculation from a boar for most of the time of high genetic value ;
- This technique saves time, as several females can be inseminated at the same time and the farrowings can be grouped together;
- by eliminating physical contact between boar and sow, this has the advantage of reducing the risk of trauma and injury during mating and minimising the risk of spreading venereal diseases.
- In preparation for AI, semen quality is monitored and assessed throughout the process. This means that infertile boars can be eliminated quickly.

In sows, two possibilities are exploited (Figure 11), namely artificial insemination at cervical level, which can be carried out with the Gédis, and artificial insemination at post-cervical level, which requires a catheter connected to a semen container.

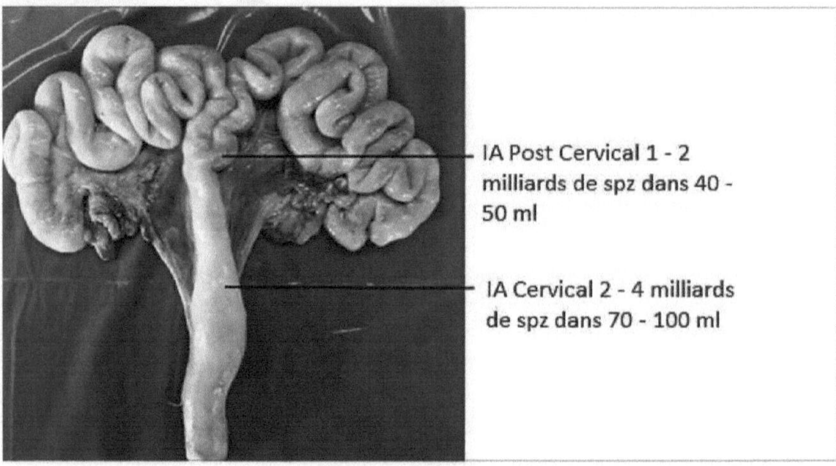

Figure 11: Female genitalia and location of AI semen (Guy delhomme, IMV Technology)

PART II: EXPERIMENTAL STUDY

Chapter I: Materials and methods

I. Hardware

1.1. Study area

The study was carried out at the CMAP biotechnology laboratory in Loumbila. The Loumbila breeding station is located 15 km from Ouagadougou on the Ouagadougou-Ziniaré road. The commune of Loumbila is bordered to the east by the commune of Ziniaré, to the west by the commune of Pabré and the arrondissement of Nongremassom, to the north by the commune of Dapelogo and to the south by the commune of Saaba. The commune of Loumbila covers an area of 362 km^2 with a population density of 65 inhabitants per km .[2]

Artificial insemination was carried out in the Ouagadougou and Bobo-Dioulasso zones. In the Ouagadougou area, average rainfall between 2009 and 2018 was 870.12 mm per year (Karambiri, 2023). The rainy season extends from May to October, and the average number of rainy days is 67 per rainy season. The average monthly temperature is around 30°C. During December and January the minimum temperature can be as low as 19°C. The maximum temperature is 40°C between April and May.

In the Bobo-Dioulasso area, average rainfall for the period 1956 to 2018 was 1,042.67 mm (Karambiri, 2023), with an average annual temperature of between 19.5°C and 36°C. Almost the entire municipality lies on a primary sandstone plateau that stretches between latitudes 11° and 13° north and the Fouta Djallon as far as the south-west of Burkina Faso.

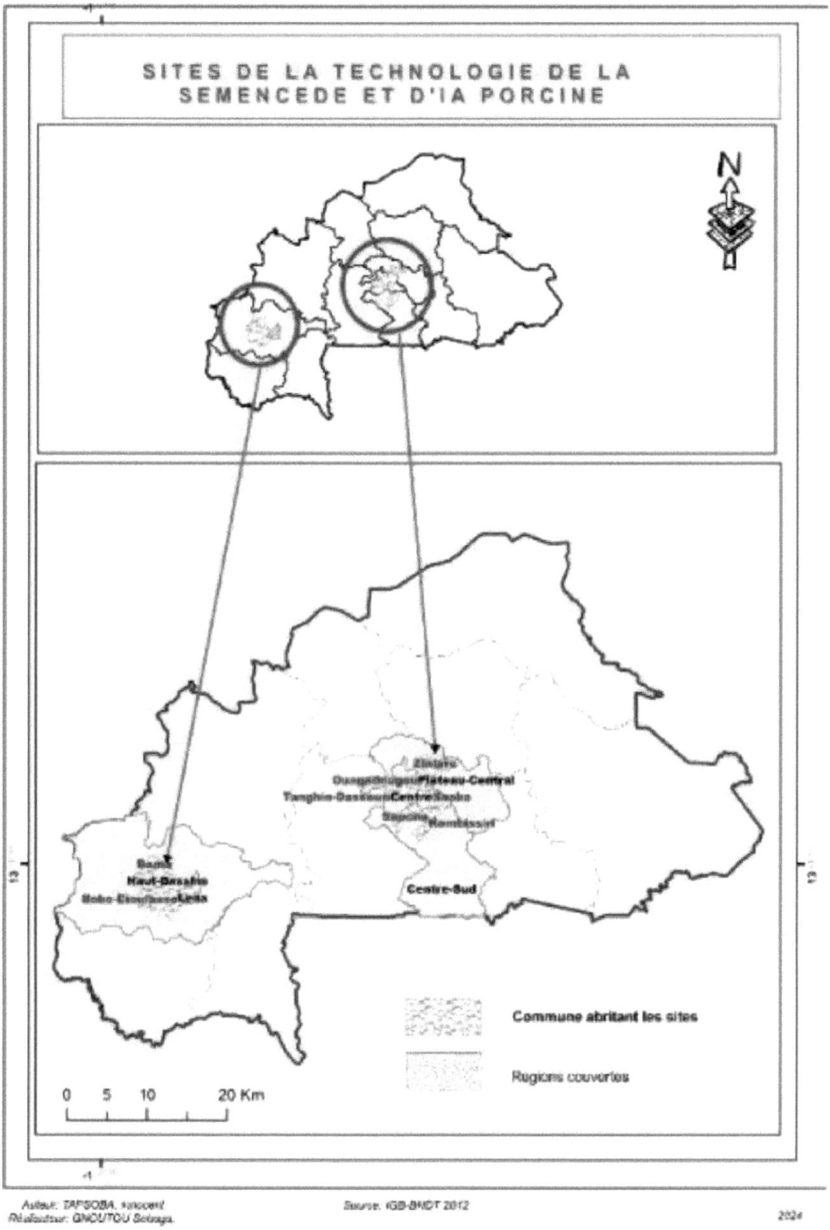

Auteur: TAPSOBA, Innocent
Réalisateur: GNOUTOU Sebaga.
Source: IGB-BNDT 2012
2024

I.2. Biological material

For semen technology, the biological material consisted of 40 ejaculates from eight (n= 08) boars, including 02 large white, 02 landrace, 02 duroc and 02 piétrain boars (Table I).

Table 1 Characteristics of boars used for seed harvesting

Breed	Large	Large	Landrace	Landrace	Duroc	Duroc	Piétrain	Piétrain
Age (months)	16	16	16	16	15	15	15	15
Weight (kg)	217	125,5	198,5	189	199	99,5	169	181
Body length	131	102	122	123	123	98	122	107
(cm)	138	111	131	132	134	107	122	126
Height at withers (cm)	90,5	78,5	83	86,5	86,5	73	84	83,5
Ear length	22,5	22,5	23	23	24	19	25,5	24
(cm)	30	26,5	31	29	28	22	27	28
Pool width	21	19,5	23,5	21	23,5	18	25	23

Artificial insemination using semen produced in Burkina Faso was tested on 145 sows distributed in the Ouagadougou and Bobo-Dioulasso areas as follows (Table II).

Table 2 Characteristics of inseminated sows by zone

Location	Sows (n=103)	Gilts (n=42)	Average age (months) Sows	Average (months) Gilts	Tower of chest (cm)	Height at the (cm)	Length the body (cm)
Ouagadougou and surrounding areas	48	37	14,6	9,6	105,2	67,7	92,3
Bobo-Dioulasso and periphery	55	5	24,2	11,6	109,6	69,6	95,8
Total/Average	103	42	18,46	10,6	106,9	69,6	95,84

I.3. Technical equipment

The technical equipment used in semen technology consists of collection equipment (a dummy, plastic bags, filters, a cup, gloves) and semen analysis equipment (an electronic balance, a microscope, a spectrophotometer).

For the artificial insemination experiment, the equipment consisted of gedis for semen packaging, wine cellars for storage and thermo-adjustable coolers for transport to the field.

There was also a lassoporc for restraint, gloves, lotus, gel and soap.

II. Methods

II.1. Seed training and harvesting

The boars were trained for four (04) weeks to harvest semen. This consisted of introducing each boar into the harvesting room where a dummy was installed. Three stimulation techniques were used depending on the boar.

Our study used a semen collection procedure similar to that proposed by Niwa et *al* (1954) cited by Anne (2002). It consists of ensuring that the corkscrew end of the penis is locked in the spiral uterine cervix of the sow by manually exerting pressure on the end of the penis.

Three stimulation techniques were used depending on the boars. Firstly, the boars were stimulated using a handkerchief soaked in the urine of sows in heat;

If they refused, these boars were injected with 0.5 to 1 ml of dinolytic. The final technique was to leave the slowest boars in last position to take advantage of the drool of the boars that were positive to the excitement. This process allowed the animals to familiarise themselves with the dummy as if they were in the presence of a sow.

After arousal and jumping onto the dummy, once the penis is out, the operator grabs it, pulls down while maintaining tension and exerting pressure on the foreskin. The semen is collected in a 1 litre beaker (Figure 12) into which a sterile plastic bag is placed. The edges of the bag are secured to the cup with an elastic band. The container is then covered with a filter that separates the tapioca that is retained and allows the seed to pass through, as described by Barrabes et *al* (2008). The reaction time to the dummy and the duration of ejaculation were recorded using a stopwatch.

For the examination of semen, especially undiluted semen, Buisson (1956) recommends warming the samples by shaking them in air for 2 hours at 37°C.

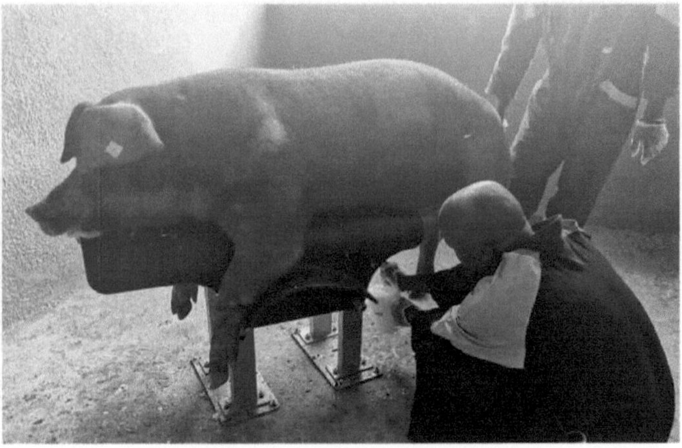

Figure 12: Seed harvesting session

II.2. Seed analysis

11.2.1. Macroscopic parameter

Macroscopic analysis focuses on volume, colour and odour using the method of Leborgne et *al.* (2013).

The volume of ejaculate was measured by direct reading using the graduations on the collection tube. The reading was taken without taking into account the foamy part of the ejaculate. Colour, viscosity and odour were assessed using the collector's sense organs (eyes, nostrils).

11.2.2. Microscopic parameters

Each ejaculate was weighed using a sensitive OHAUS adventurer balance, model PX4201, capacity 4200g, accuracy mg of 100, serial number C05174147, previously tared with the collection container. The volume in ml is obtained after rectification with the density value of the boar semen.

The sperm concentration in the ejaculate was obtained using a spectrophotometer and confirmed by a MetroSperm.

Microscopic examination at magnification 100 of a drop of pure seed between slide and coverslip on a hot plate at 37°C was used to assess mass and individual mobility and vivacity.

II.3. Packaging and storage

After analysis, the natural semen was diluted with a TRIXcell Ultra dilutor and distilled water.

The concentration (in millions/ml) of the ejaculate is multiplied by the volume (in ml) of the ejaculate to obtain the total concentration (in billions). The total concentration is divided by 3 billion to obtain the number of doses to be produced. The number of doses is multiplied by 90 ml to obtain the final volume and finally the volume of the ejaculate is subtracted to obtain the volume of the dilutor. The diluted semen obtained is then packaged in gedis (Figure13). Packaging is carried out using a "Golden Field" bagging machine. Each dose has a volume of 90 ml and contains 3 billion spermatozoa. The sachets are vacuum-filled at room temperature (25°C) and stored in a wine cellar at between 17 and 19°C.

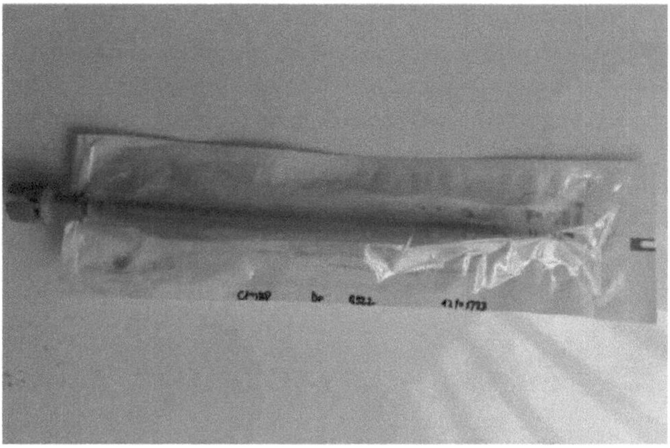

Figure 13: Gédis

The diluted seed is checked daily over a period of 07 days to assess how long the seed produced is valid and can be used.

II.4. Preparing for artificial insemination

II.4.1. Induction of oestrus

Artificial insemination initially consisted of identifying the lactating females that could be selected. Stakeholders were asked to allow the piglets to suckle for 28 days (Memento, 2006) before weaning them, which is a factor that triggers the expression of heat in sows. Thus, for the Ouagadougou zone and surrounding area, weaning was carried out on 20 April 2023 and for the Bobo-Dioulasso zone on 27 April 2023. To these sows, we added those whose return to heat could coincide with the insemination period or gilts of breeding age that would come into heat during the operation.

II.4.2. Insemination proper

Insemination was carried out on pre-selected sows and gilts showing heat. The artificial insemination operation involved 145 sows aged 18 months on average, from 25 farms, with a live weight of between 74 kg and 194 kg, compared with an average weight of 172 kg for the boars harvested (99.5 < X < 217 kg). However, 144 were selected for the study.

Most of the standing females were restrained using a pig lasso or simply by the ears and tail (Figure 14), depending on docility. The vulva was sprayed with a disinfectant solution followed by manual cleaning with toilet paper to remove any foreign bodies. After cleaning the vulva, the Gedis is inserted into the vagina over at least 4/5ème of its length and left for two (2) to five (5) minutes, the time required for total absorption of the semen. The gedis is a probe that simultaneously stores and inseminates the sow. The head of the probe has fins that are placed in the cervix, like the foam head or the spiral head, and the semen reserve, stored under vacuum in a membrane around the stem, is placed in the vestibule and vagina. Under the effect of body temperature, the wax plug obstructing the probe head liquefies at around 37°C, two to three minutes after insertion, and vaginal contractions then release the semen (Feller et *al.,* 2004). A five-minute wait ensures that the semen reaches its destination. Finally, the catheter is removed from the vagina to be disposed of or destroyed. We used 69 doses of duroc, 43 doses of large white and 32 doses of landrace.

Figure 14: Insemination of a sow

II.5. Pregnancy diagnosis or pregnancy report

After insemination, gestation was checked by observation of non-returns to heat (21 days after AI) and by ultrasound (35 days after AI). The ultrasound method has a definite advantage in that, in addition to checking gestation, it allows the number of intrauterine pups, the sex of the foetuses and the variability of the foetuses to be verified (Hanzen *et al,* 1992). The model of ultrasound scanner used for the present study is a KAIXIN, equipped with a sectorial probe with a frequency of 50 - 60 Hz ± 1. The operation consisted of applying gel to the tip of the probe and placing it above the inguinal udder (Figure 15), in the groin fold of the sow in restraint.

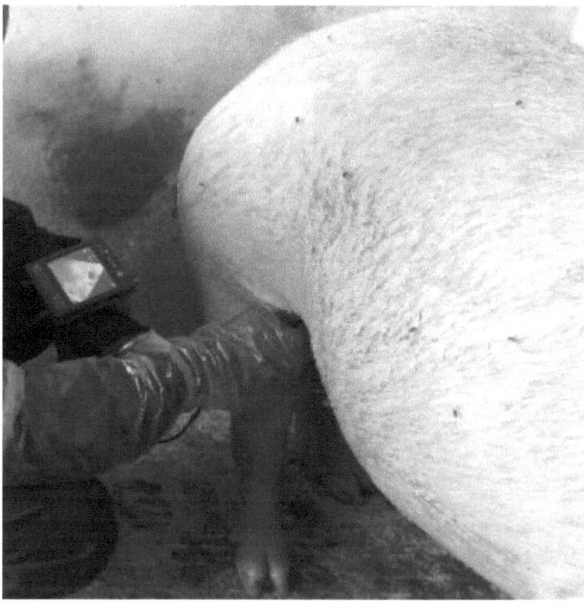

Figure 15: Ultrasound scan of a sow

II.6. Statistical analysis

Data on semen characteristics and artificial insemination results were recorded using Excel software. For semen technology, statistical analyses were carried out on semen characterisation, concentrations and mobilities according to breed. For inseminations, the analysis of variance (ANOVA) focused on the fertility rate as a function of breed and individual. The Pearson correlation test was used to assess relationships between variables using XLSTAT 2014.5.03 software. Variations according to boar breed are expressed as mean ± standard deviation and differences between means are considered significant at the probability threshold of $P < 0.05$.

Chapter II: Results and discussion

I. Results

1.1. Boar training and semen characteristics

The boars were trained to harvest seed for 28 days at a rate of two harvests per session per week. The acceptance rate at jumping or training was 7/8 (87.5%). The pedigree boars responded positively to the training but their semen was small (30-50 ml), clear and sperm-free.

Table III below shows the jumping behaviour and characteristics of the ejaculates.

Table 3: Characteristics of the ejaculates collected

Variables	Average General	Large white	Landrace	Duroc
Vol_ejac (ml)	171,68	182,66±58,76a	191,08±0b	141,3±14,59c
Conc ejac (million/ml)	277.37	230,19±11,13a	195,45±0b	406,48±73,75c
Number of doses	19,72	15,7±5,83a	12,5±0b	22,87±7,07c
Motil (0 - 5)	2,99	2,97±0,06a	3±0b	3±0b

A total of 33 ejaculates were collected with a total volume of 4296.5 ml. The average volume of the ejaculates was 171.68 ml. The highest average was recorded in landrace with 191.08 ml and the lowest in duroc with 141.3±14.59 ml. The average concentration was around 277.37 million sperm per ml; the duroc had the highest concentration (406.48±73.75 million/ml), followed by the large white (230.19±11 million/ml) and finally the landrace (195.45 million/ml). The number of doses being equal to the concentration times volume of the ejaculate divided by 3 billion justifies therefore that the duroc in spite of the small volume of ejaculate ends up with an average of 22.87 doses against the landrace which had the largest volume but totalled an average of 12 doses.

I.2. Packaging and advice

Semen from each boar after motility evaluation on day one was diluted and stored for 07 days. Figure 16 below shows the variation in sperm motility for each breed during the first week of storage.

The curves for duroc and landrace overlap.

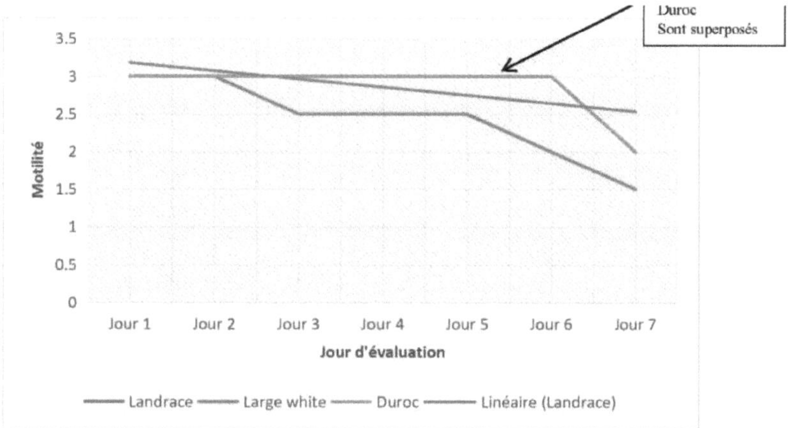

Figure 16: Changes in Spz motility by breed during storage on a scale of 0 to 5

Irrespective of the individual and breed, motility was greater than or equal to 2.5 (50%) until day 05ème of storage, before falling from 2.5 to 1.5 between day 05ème and day 07ème .

For libido expression in boars, the mean reaction time was 3.13 ± 0.06 minutes for large white, 3.08 ± 0.00 minutes for landrace and 4.09 ± 0.00 minutes for duroc. The mean duration of ejaculation was 5.59 minutes, 5.75 minutes and 5.00 minutes for large white, landrace and duroc respectively. Semen colour was always creamy white.

I.2. Artificial insemination and pregnancy tests

Only females expressing heat after weaning, or sows and gilts expressing heat during the period of the operation on the farms, were inseminated.

I.2.1. Post-parturition parameters

The sows selected from 25 farms varied greatly. Piglet suckling time averaged 55 days, with extremes of 32 days for early weaners and 94 days for late weaners, as shown in Figure 17 below. The time from last farrowing to AI ranged from 17 to 67 days. The time between weaning and fertilizing heat was 2 to 23 days, and the time between the first signs of heat and the first insemination was 24.81 to 37.5 hours.

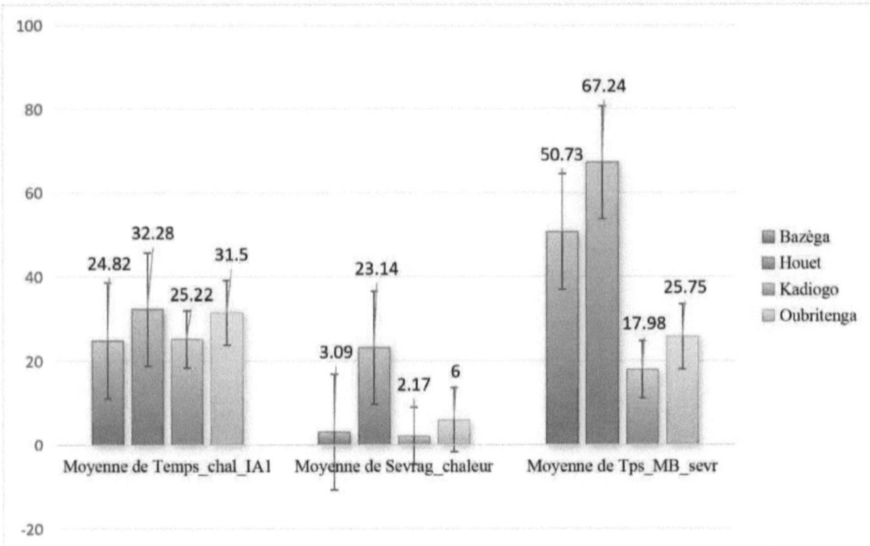

Figure 17: Characteristics of calving-related parameters

1.2.2. Heat characteristics and artificial insemination

In this study, signs of swelling of the vulva, redness of the vulva, agitation, erect ears, secretion of mucus, sniffing followed by jumping were observed in females in heat. The interval between the first signs of heat and the first insemination was 24 hours. The second insemination was carried out 12 hours after the first, i.e. 36 hours after the first signs of heat, and was only possible on a few sows due to the constraints of time and distance between farms.

1.2.3. Gestation rate

After artificial insemination, pregnancy was established by observing the non-return of heat at 21ème days after AI and confirmed by ultrasound examination between 35 and 45 days after AI. Of the 145 females inseminated, one was discarded because she was already pregnant during the operation and was inseminated by mistake (Figure 18).

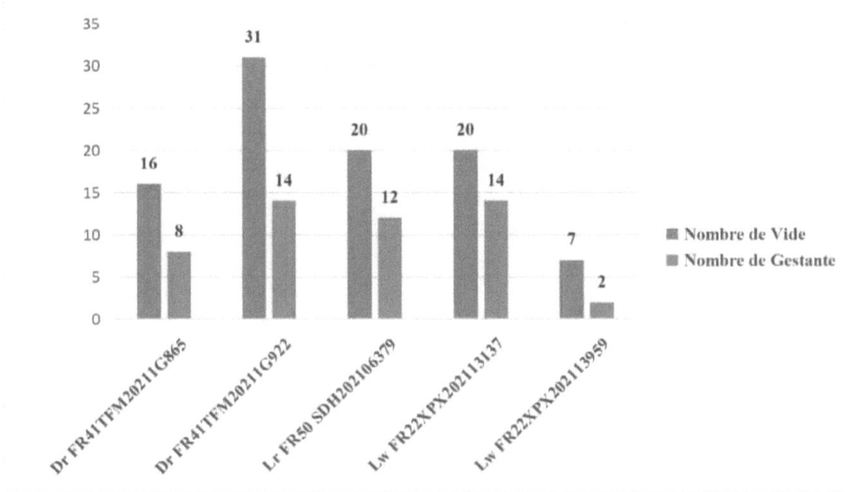

Figure 18: Monitoring of gestation per boar

The results for non-return of heat were 83/144, i.e. 57.63%. Ultrasound scans were performed on all the sows inseminated and only 51/144 sows were truly pregnant, i.e. a pregnancy rate of 35.42% compared with the initial number of sows (n=144). Of the 51 pregnancies, the pregnancy rate by breed was 45.10%, 31.37% and 23.53% for Duroc, Large White and Landrace respectively.

I.2.4. Gestation period

The average gestation period was 113.77 ±2.66 days, with extremes of 101

days for the earliest gestation 119 days for the longest (figure 19).

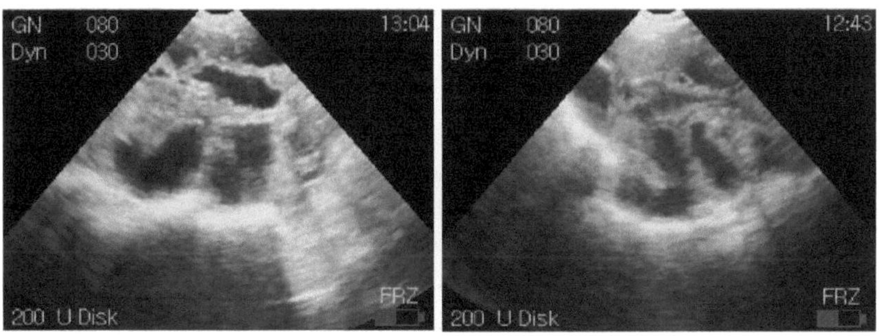

Figure 19: Ultrasound images of a 35-day gestation

(NB: The black spots correspond to the sections of the embryonic vesicles)

I.3 Scope characteristics

Table 4: Effect of boar breed on litter characteristics

Breed	Number of piglets	Average litter weight (kg)	Born alive
Largewhite	9,11±3,72a	1,31±0,29a	8,05±3,19a
Landrace	9,4±3,27a	1,64±0,26b	8,9±3,38a
Duroc	8,82±3,08a	1,32±0,33a	8,29±3,21a
P-alue	**0,9812**	**0,0291**	**0,8667**

The total number of piglets born from the AI operation was 463. The mean litter size was 9.07 ± 3.32 piglets. The mean birth weight of the piglets was 1.40 kg ± 0.33 and the mean number of piglets born alive was 8.34 ± 3.18. The mean weight of landrace piglets was 1.64±0.26 kg compared with 1.32±0.33 kg and 1.31±0.29 kg for duroc and large white piglets.

The number of piglets born per sow was not correlated with the sow's farrowing rank. However, the number of piglets born alive and litter size, as well as the age of the sow and the number of farrowings recorded, were strongly correlated (0.80 and 0.90 respectively).

II. Discussion

The aim of introducing boars to the Centre National de Multiplication des Animaux Performants was to study the possibility of producing fertile boar semen in Burkina Faso from imported purebred animals. 87.5% of boars responded favourably to seed collection. However, 25% of this rate had ejaculate volumes of less than 50 ml and clear semen without spermatozoa. This difficulty could be linked to the collection period, which coincided with the high temperatures of the year. Contrary to the ideal temperature of 15°C to 20°C recommended for these exotic breeds according to Feller et *al* (2004), the average temperature recorded was 30.68°C. Anne (2002) speaks of sterility due to summer temperatures. Added to this is the fact that no analysis has been made of testicular anomalies, infections or tumours.

Le Roux (2002) also considers that bacterial contamination of semen is inevitable and systematic, and the consequences for semen quality depend on the type of dominant bacteria and the bacterial load of the ejaculate. Althouse et *al* (2000) have even succeeded in demonstrating the presence of spermicidal bacterial germs in semen, showing a reduction in mobility and agglutination of the spermatozoa, with death of the spermatozoa after two days' storage. But despite all these factors influencing the quality of the semen, motility remained above 60% until day 5ème .

Regarding ejaculate volume, our results were 182.66 ml, 191.08 ml and 141.3 (ml) respectively in large white, landrace and duroc boars. These results are higher than those obtained by Dotché et *al* (2019). However, they are still lower than the average of 300 ml found by Chamoulaud

(1995), which is a result to be qualified because the volume of tapioca was on average 115 ml. In our study, as in Dotché's, the gelatinous "tapioca" fraction was not included in the volume of the ejaculate.

According to Guillouet et *al* (1999), the volume of ejaculate increases from 9 months to 15 months of age when it stabilises. Thus, the delay in training and the start of boar semen harvesting did not allow a good realisation of the volume of semen production of our boars despite their average age of 15 months.

The mean concentration (x 10^6 spz/ml) of the semen of the large white boars was higher than that of Elile et *al.* (2014) in Nigeria which is 187.8 and that of Thema et *al.* (2022) which is 120±39. However, it is lower than that obtained by Guillouet et *al.* (1999), which is 321± 89. The concentration of landrace seed was lower than that of Guillouet et *al.* (1999), which was 279±99. Duroc had a higher concentration than broad white and landrace, whose environmental adaptation seemed better.

With regard to storage time, our study chose to store fresh semen produced with TRIXcell Ultra as a diluent. The choice of TRIcell Ultra was based on the fact that the motility observed was 78.6± 2.4, 10 days after harvesting fresh seed and storage at 17°C+1 (Van Den Berg et *al.*, 2014). These results are higher than ours, which were 60% and 5 days of storage. Van Den Berg et *al* (2014) believe that TRIcell compared to Beltsville Thawing Solution (BTS) results in more piglets being born alive and this is partly related to its broad spectrum combination of three antibiotics.

Our pregnancy success rate of 35.42%, made up of 45.10%, 31.37% and 23.53% for Duroc, Large White and Landrace respectively, is much lower than that obtained following a double insemination by Bitty (2014) in Côte d'Ivoire, which was 69.44% using imported semen. However, the rate is slightly higher than the first pig artificial insemination test carried out by the Ministry of Animal Resources and Fisheries in 2021, which recorded a success rate of 33.33% with imported semen, on a population of 36 sows (MARAH, 2021). The rate we recorded was low compared with the success rate in pig AI in developed countries, which is over 90% (IFIP, 2013). This could be explained by the diversity of farms and their management, the diversity of animals, the length of storage and the quality of the semen. The success of AI is not simply a question of biotechnology but of the physiological state of the sows, which is an important factor in success (Bariteau, 1977).

Litter size is higher than that of the local breed obtained by Youssao (2009) which is 7.5 ± 1.91; but lower than that of Dourmad et *al.* (2010) which is 14.0 for total piglets born and seems corroborated with FAO (2012) which is 7 to 15 piglets per farrowing per sow. However, the 13.7% live-born piglet loss rate found by Dourmad *et al* (2010) is similar to our results. The average number of piglets born alive is lower than that recorded by Uzu et *al.* (1979) which was 9.47 with a range of 1.55 piglets between the extreme boars.

The increase in the number of foetuses is accompanied by an increase in the total flow of blood to the uterus but also by a reduction in the flow per foetus and, consequently, in the supply of nutrients (Pere, 2000), which should logically lead to a reduction in the birth weight of piglets. This is confirmed by analysis of the influence of litter size on piglet birth weight (Quesnel et *al.* 2008), indicating that there are more small-weight piglets (< 1.0 kg) and fewer heavy-weight piglets (> 1.8 kg) in larger litters. This analysis is in line with the results we obtained, where the average weight is 1.4 kg ± 0.33 but higher than the local breed, which is 828 g for males and 731 g for females (Youssao, 2009). So it is not because the number of piglets from a farrowing is reduced that we will automatically have large piglets. Canario et *al* (2007) estimated that for the same litter size, piglets produced by current sows are heavier at birth than piglets produced by sows in 1977.

CONCLUSION AND OUTLOOK

The study on pig semen technology shows that boar semen production is feasible in Burkina Faso.

The adaptation of piglets at the end of post-weaning to the change in diet and rearing conditions, and the production of fertile seed, reassure us that these breeds can be introduced directly to improve productivity on farms.

The rate of boars responding to the experiment, the average volume of ejaculates and the concentration in million spermatozoa per millilitre reached the standards required for bagging/dosing. The doses were transported for artificial insemination in strict compliance with the recommended temperature of 17°C.

In the field, insemination was carried out without the slightest change in the producers' breeding practices. Dissemination of the semen produced in this study by artificial insemination provided an opportunity to assess the fertility of the boars. The adaptation of the duroc breed appears to be better than that of the landrace and the large white, whose cross-breeds are the best known and considered to be the best adapted exotic pig breeds. The study also showed that the piétrain seems to have difficulty adapting to the results obtained.

Pig semen technology will make it easier to adopt artificial insemination for pigs by making boar semen available and accessible, and to improve the productivity of pigs in Burkina Faso.

Compared with Western countries, the success rate is low, but this could be explained by the poor control and lack of rigour in heat detection and the right time for artificial insemination.

This study opens up new perspectives, such as an understanding of follicular dynamics in the local breed; a comparative analysis of the success rates of cervical (Gédis) and post-cervical (catheter probe) insemination in the local breed; and the fertility of boars for semen collection and sows for artificial insemination;

All in all, pig semen technology is contributing to the development of the pig industry in Burkina Faso, and the Duroc breed is better suited to the task.

Bibliography

Agbokounou A. M, Gbênangnon S. A, Issaka Y. A, Karim, Guy A. M, Bénoît K., Jean-Luc H, 2016. Characteristics of local pig breeding in Africa, Journal of Animal and Plant Sciences, Vol.30, Issue 1: 4701-4713.

Althouseg.C., Kuster C.E., Clark S.G., Weisiger R.M, 2000. Field investigantion of bacterial contaminants and their effects on extended porcine semen. Theriogenology, p 11671176.

Annick Audiot, 1995. Races d'hier pour l'élevage de demain. INRA Paris, 229p.

Barrabes A. S., Gary B. G., Bouvier B. P., 2008. Colllectis R automatic boar collection technology. Theriogenology 70, p 1368-1373

Barone R., 2000. Anatomie comparée des Mammifères Domestiques. éd. Vigot, 4, 238-247.

Bariteau F., Bussiere J., Courot M., 1977. Insémination Artificielle porcine : Améliorations techniques, résultats recents. J.R.P. en France, p 11-14

Bitty Z.B.A., 2014. *Evaluation of the effectiveness of artificial insemination in pig genetic improvement in Côte d'Ivoire.* Master's thesis in animal production and sustainable development. Cheikh Anta Diop University of Dakar. EISMV de Dakar. 43p.

Bories P. C. E., 2009. *Chronopart de la truie en élevage : effet de la durée de mise bas sur quelques paramètres physiologiques et biochimiques.* Thesis, Veterinary Medicine, Toulouse 3, 112p

Boulot S., Badouard B., 2008. Changes in insemination practices in French sow herds from 2001 to 2006. Journ. Rech. Porcine Fr. 40, 257-258.

Brassard P., Martineau R. and Twagiramumgu H., 1997. Fixed time insemination finally possible. Dairy cattle symposium (conference), Canada, 1997, CPAQ, 16 p.

Canario L., Rydhmer L., Gogué J., Tribout T., Bidanel J.P., 2007a. Estimation of genetic trends from 1977 to 1998 for farrowing characteristics in the French Large White breed using frozen semen. Animal, 1, 929-938.

Da M., 2004. *Monitoring cyclicity in sows.* Engineering thesis, Institut du Développement Rural/Université Polytechnique de Bobo-Dioulasso (IDR/UPB) 79 p.

Dacheux J.L., Paquignon M., 1987. Epididymal maturation of spermatozoa, influence on gamete quality. *J.R.P en France,* 19, 99-106.

D'Orgeval R, 1997. *Le développement de l'élevage porcin en Afrique : l'analyse des systèmes d'élevage du porc local africain au Sud Bénin.* Doctoral thesis from the Institut National Agronomique, Paris Grignon, 273p.

Dotché I O, Kiki P, Govoeyi B, 2019. Status report on animal artificial insemination in West African countries. Journal of Applied Biosciences 143: 14712 - 14730.

Dourmad J-Y, Canario L L., Gilbert H, Merlot E., and Quesnel H, 2010. Evolution of animal performance and robustness in pig farming. INRA Productions Animales. P.53-64.

Du Buisson F. D. M. 1956. Conservation of boar semen after dilution and maintenance of its fertilizing power. Annales de zootechnie, III, 5 (3), p.195-212.

Elile FC, Ogbu CC, Ugwu SOC, Machebe NS, 2014. Libido and ejaculate characteristics of boars exposed to direct solar radiation. The Journal of Animal and Plant Sciences 24: 43-49.

FAO, 2012. Pig sector, Burkina Faso. National Livestock Reviews of the FAO Animal Production and Health Division. No. 1 - Rome, FAO. - 83p.

Feller D., Thilmant P., Wavreille J. and Boudry C., 2004. Le Verrat, la Truie aspects technique de la reproduction. *Filière porcine wallonne,* n° D/2005/8887/1, 48 p

Flowers B., 1996. Reproductive physiology of the boar. Proceedings of 29th Am. Meet. Of Association of Swine Reproduction, Desmoines, Iowa, 1-6.

Geneviève J., 2019. Use of new biotechnology tools for xenotransplantation. In: Bulletin de l'Académie Vétérinaire de France tome 1. pp. 89-93

Guillouet P., Tribout T., Bussière J.F., BertaudG., Bidanel J.P., Terqui M., 1999. Analysis of variation factors in sperm production in artificially inseminated boars. Journ. Rech. Porcine Fr, 31, 45-52.

Guy Bernard, D.T.A, 2003. New technologies in artificial insemination. Centre d'insémination porcine du Québec, Porc Québec 7p.

Hanzen C, Laurent Y, Jakovljevic S, 1993. Application of ultrasound in bovine reproduction 2. The pregnant and non-pregnant uterus. Ann. Méd. 137 : 93 - 101.

Heape W.,1900. The "Sexual Season" of Mammals and the Relation of the "Pro-oestrum" to Menstruation Quarterly Journal of Microscopical Science. New Series. 44 : 1-70.

Heath E., Olsen A., Basurto-Kuba V. 1984. Intratesticular spermiostasis in the boar. 10th Congr. Anim. Reprod. and A.I., Illinois USA, III, 272-273.

Holst S.J., Sterility in boars. Nord. Vet. Med, 1949, 1, 87-120

IFIP, 2013. Mémento de l'éleveur du porc. IFIP (Editor), 7th edition, Paris, 364pp.

ITP, 1991. Mémento de l'éleveur de porc, fourth edition, 566 p

Karambiri B.L.C.N, Gansaonre R.N, 2023. Spatio-temporal variability of rainfall in the climatic zones of Burkina Faso: the case of Bobo Dioulasso, Ouagadougou and Dori. European Scientific Journal, ESJ, 19 (9), 262.

Kiendrébéogo, T., 2006. *Diagnostic des élevages porcins dans la zone de Bobo Dioulasso système d'élevage et conditions technico-économiques de production. Bobo Dioulasso*: DEA dissertation at the Institut de Développement Rural (lDR) de J'Université Nazi Bonî (ex-Université Polytechnique de Bobo-Dioulasso (UPB). 71p.

Ladds P.W., 1996. The Male Reproductive System. In: SIMS L.D. and glastonsbury J.R.W. Pathology of the Pig. Victoria, 407-418.

Leborgne M.-C., Tanguy J.-M., Foisseau J.-M., Selin I., Vergonzanne G., Wimmer E., 2013. Reproduction des animaux d'élevage. 3ème. Paris, 467p.

Le Roux A. N., 2002. *La réforme des verrats de centres d'insémination artificielle pour baisse de qualité de semence : approche anatomopathologique et histologique.* Toulouse National Veterinary School, Thesis, 140p.

Lukusa B. D, 2005. *Amélioration de la pratique d'élevage de porcs dans les milieux urbains et périurbains de la ville de Kinshasa.* Dissertation for the Diplôme d'Études Spécialisées en Gestion des Ressources Animales et Végétales en Milieux Tropicaux Faculté Universitaire des Sciences Agronomiques de Gembloux, ULG, 50p.

MRA, 2003. Diagnostic de la filière porcine au Burkina, PAPISE, 78p.

MRA, UNDP 2011. Contribution of livestock farming to the economy and the fight against poverty, the determinants of its development, 78p.

MARAH, 2021, report Bilan annuel 2021 des activités du ministère..). - Vet. Rec. 68, 6-267, 169.

MARAH, 2021. Enquête nationale sur le cheptel (ENC1) 2018, analysis report, 89p.

Martin R. S., Martinez E., Garcia-Artigac, De Alba C., 1996. Boar Semen Evaluation in Practise. Reprod. Dom. Anim, 31, 519-526.

Martin Y.C.M., 1993. Quality of semen for A.I. pigs. Update of a colorimetric test. Th : Med. Vet. Toulouse : -TOU 3, 4032.

Mermillod P., Guillouet P., Berthelot F., Martinat-Botté F., Plat M., Terqui M., 2003. Biotechnologies for pig reproduction. Journées Recherche Porcine, 35, 323-338.

Mémento de l'agronome, 2006. CIRAD - GRET, Ministry of Foreign Affairs, 1691p.

Nonfon WR, 2005. *La filière de production du porc local au Bénin : l'amélioration de sa productivité par l'alimentation.* PhD thesis in Agronomic Sciences and Biological Engineering, Faculté Universitaire des Sciences Agronomiques de Gembloux, Belgium, 236p.

Paquignon M., Bussiere J., Bariteau F., 1988. Efficiency of boar semen conservation techniques. La revue INRAE. Productions Animales. Vol1.N°4.

Paquignon M., Bussière J., Bariteau F. 1988. Efficiency of boar semen conservation techniques. Productions Animales, 1 (4), p.271-280.

Père M.C., Etienne M., 2000. Uterine blood flow in sows: effects of pregnancy stage and litter size. Reprod. Nutr. Dev, 40, 369-382.

Polge C., 1956. Artificial insemination in Pigs. Vet. Rec, Jan: 62-76.

Quesnel H., Brossard L., Valancogne A., Quiniou N., 2008. Influence of some sow characteristics on within-litter variation of piglet birth weight. Animal, 2, 1842-1849.

Signoret J.P., 1972. *Contribution à l'étude des mécanismes éthologiques et endocriniens du comportement sexuel de la truie.* Doctoral thesis. University of Paris VI. CNRSA.O., 7257, 180p.

Thema, M. A., Mphaphathi, M. L., Ledwaba, M. R., & Nedambale, T. L. (2022). Investigation of the efficacy of different semen extenders and in vitro storage period on Windsnyer boar sperm quality equilibrated at 18°C. Reproduction in Domestic Animals, 57, 1644-1648. https://doi. org/10.1111/rda.14256

Toussaint M., 2000. Maîtrise du début de l 'œstrus chez la truie de race locale : Approche endocrinologie, cytologique et mesures du pH. Thesis, Cheikh Anta Diop University, Dakar, Interstate School of Veterinary Science and Medicine, 139 p.

Uzu G., Bonneau M, J Lebost, 1979. Influence of boars on the main parameters of herd productivity and on the length of gestation. Annales de zootechnie, 28 (3), pp.315-323.

Van den Berg B.M., Reesink J, and Reesink W., 2014. TRIX cell+, a new long-term boar semen extender containing whey protein with higher preservation capacity and litter size. Open vet. 4 : 20 - 25.

Vaissaire J.P., 1977. Sexuality and reproduction of domestic and laboratory mammals. *Maloine-S, A. éditeur-Paris France,* p 157-267

Vincent D., 2014. Anatomy of the genital and udder systems, 6 p

Wrobel K.H. 1990. Male Reproductive System In: DELLMAN. Textbook of Veterinary Histology, 2nd edition, 226-243.

Youssao A.K.I., Koutinhouin G.B., Kpodekon T.M., Yacoubou A., Bonou A.G., Adjakpa A., AHOUNOU S., TAIWO R., 2009. Genetic improvement of the zootechnical performance of the local pig in Benin by crossing with the Large White.

OECD/FAO (2022), OECD and FAO Agricultural Outlook; OECD Agricultural Statistics (database), https://doi.org/10.1787/agr-outl-data-fr.

APPENDICES

Photos of the verrassons 3 days after their arrival (15/03/2022_CMAP_Loumbila)

Pig farm

Ultrasound

First dose of boar semen produced at BF on 23/03/2023 at Loumbila

Freshly harvested seed icebox for conservation of doses

A: Dummy, B: Collection bag; C: Balance; D: Spectrophotometer; E: MetroSperm F: Microscope; G: Oven; H: Water bath

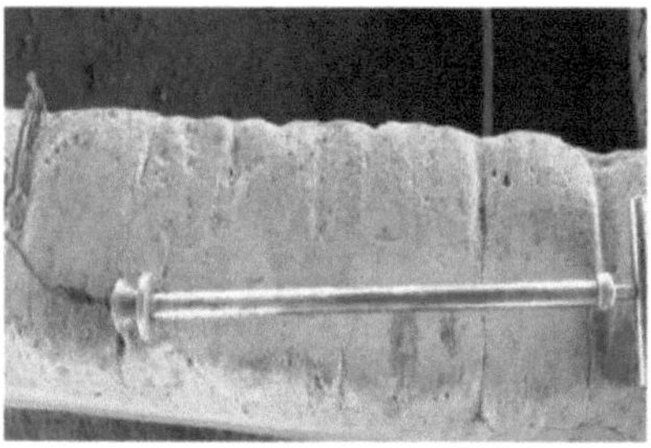

Pork lasso

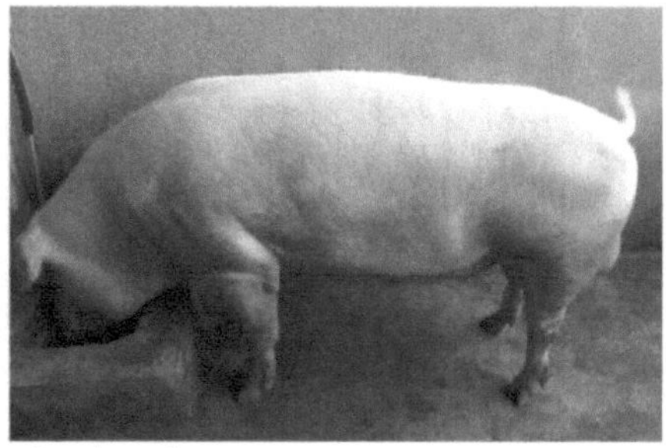

Landrace boar

Verrat DurocVerrat

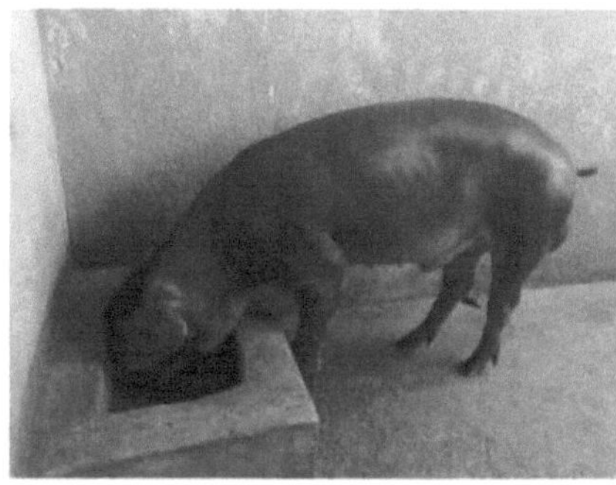

Duroc

Pietrain

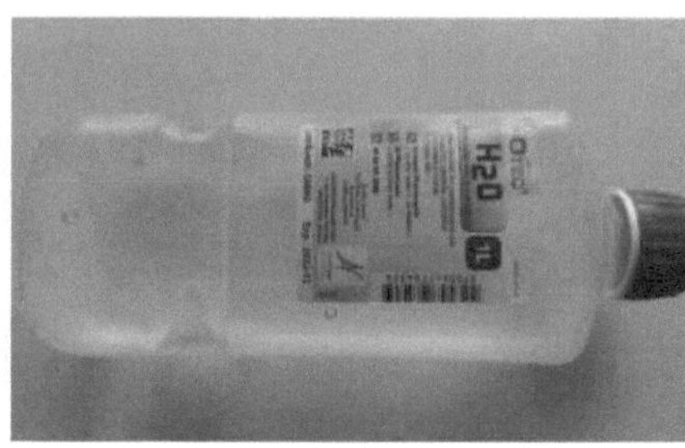

boarDistilled water

MIX
Papier aus verantwortungsvollen Quellen
Paper from responsible sources
FSC® C105338

Printed by Books on Demand GmbH, Norderstedt / Germany